SATURN

BLASTOFF!

SATURN

by Tanya Lee Stone

BENCHMARK BOOKS

MARSHALL CAVENDISH

NEW YORK

For Jacob—I hope you never stop asking "Why?"

With special thanks to Roy A. Gallant, Southworth Planetarium, University of Southern
Maine, for his careful review of the manuscript.

Benchmark Books
99 White Plains Road
Tarrytown, NY 10591-9001
www.marshallcavendish.com

Library of Congress Cataloging-in-Publication Data
Stone, Tanya Lee.
Saturn / Tanya Lee Stone.
p. cm. — (Blastoff!)
Includes bibliographical references and index.
ISBN 0-7614-1234-4
1. Saturn (Planet)—Juvenile literature. [1. Saturn (Planet)] I. Title. II. Series.
QB671 .S76 2001   00-052966

Printed in Italy

1 3 5 6 4 2

Photo research by Anne Burns Images
Cover photo:: David Ducros/Photo Researchers, Inc.

The photographs in this book are used by permission and through the courtesy of:
NASA: 7, 27, 34, 40, 49, 51. JPL: 8, 18, 25, 31, 35, 44, 45. Corbis: 10, 12.
Photo Researchers: 17, ASP/Science Source: 15, Lynette Cook/Science Photo Library: 22,
Chris Butler/Science Photo Library: 36, Davis Ducros/Photo Science Library: 42, European
Space Agency/Science Photo Library: 47. Galaxy Picture Library: 20, 28, JPL/Galaxy: 32, 54.

Book design by Clair Moritz-Magnesio

CONTENTS

1

ALL EYES ON SATURN

Hundreds of years before the Space Age began in 1957, people had no telescopes. They relied just on their eyes to observe the Universe. As they gazed into the darkened depths of space, some wondered about a few stars that seemed to move among the others. They came to be called wandering stars and were named planets, from the Greek word for wanderer.

These objects in the sky fascinated the earliest astronomers. One of them was Saturn, the second-largest planet in the Solar System. Saturn is so bright in the night sky that you can see it from Earth without the use of a telescope.

The invention of the telescope in the early 1600s, however, changed astronomy forever. It began centuries of detailed observations that would bring us closer to understanding the mysteries of our Solar System.

GALILEO, HUYGENS, AND CASSINI

In 1610, Italian astronomer Galileo Galilei became the first person to view Saturn through a telescope. He had built the instrument himself. Observing the planet, he was puzzled by what he saw—two bulges

The second-largest planet in the Solar System, Saturn has been the object of fascination for thousands of years. The Mesopotamians left the first written record of the planet around 650 B.C.

Tiny specks of light, the moons Tethys and Dione pursue their orbits around the sixth planet. They were just two of Giovanni Cassini's Saturnian discoveries. Evident among the planet's rings is the gap that bears his name.

on either side of the massive object. Galileo wrote that Saturn seemed to have "ears."

Galileo observed Saturn again in 1612. This time, he did not see the "ears." So what were these bulges he had seen, and where had they gone? No one would know the answer for nearly fifty years.

By that time, telescopes had advanced, and astronomers could see objects in much greater detail. A Dutch astronomer named Christiaan Huygens was continuously tinkering with his telescope to improve its quality. Through its lenses, in 1655, he discovered Titan, a satellite, or moon, of Saturn.

One year later, in 1656, Huygens was also able to solve Galileo's riddle. He clearly saw that Saturn did not, in fact, have "ears." Instead, Huygens observed that there was a ring around the planet. This was an incredible discovery, as scientists had never before heard of a planet having a ring. They were eager to know more.

A French-Italian astronomer named Giovanni Domenico Cassini took up the ring challenge in 1675. He observed that it was not a single ring at all. Cassini saw two rings, as well as the gap between them. In his honor, the gap was named the Cassini Division. But Cassini did not stop there. He also discovered four additional moons of Saturn—Iapetus, Rhea, Dione, and Tethys.

MORE SATURN SATELLITES

As time went on, other astronomers recorded more Saturnian satellites. William Herschel was a musician, but he had a passion for astronomy. He struggled to learn the art of lens grinding and telescope making. Eventually he had fashioned the best telescope of his time. With it, this unknown organist found fame in 1781 when he discovered the planet Uranus. Eight years later, Herschel built the largest telescope to date—40 feet (12 m) long—and identified two more of Saturn's satellites. They were named Mimas and Enceladus.

In 1898, American astronomer, William Pickering, identified the satellite Phoebe. Fifty years earlier, fellow American George Bond had discovered another of Saturn's moons—Hyperion. Bond was the first to suggest that Saturn's rings were not solid.

Photographing Saturn from an observatory in Peru, William Pickering discovered Phoebe in 1898. The farthest of Saturn's moons, Phoebe takes 550 days to orbit the planet.

The last satellite of Saturn to be discovered from Earth was Janus. This moon was located in 1966 by French astronomer Audouin Dollfus. But there was a limit to the information Earth-bound astronomers could gather. In less than ten years, the United States would launch a mission to Saturn. Through this new phase of space exploration, scientists would become much better acquainted with the ringed planet.

Seasons

Each season on Saturn lasts more than seven Earth years.

2

INITIAL MISSIONS

The word *pioneer* means the first to explore something. In the case of the *Pioneer* missions that the National Aeronautics and Space Administration (NASA) sent into orbit, these spacecraft became the first ever to travel to the outer planets. *Pioneer 10* flew to Jupiter in 1972. *Pioneer 11* followed in 1973 and then continued on to Saturn. What NASA learned from the *Pioneer* missions helped them plan the next trip taken by the *Voyager* spacecraft.

PIONEER 11

Pioneer 11 was launched in 1973 from Cape Canaveral, Florida. Successfully completing the first part of its mission, it sent images of Jupiter back to Earth. It then used the gravity of Jupiter to gain speed and catapult itself toward Saturn. It reached Saturn in September 1979, flying by the planet at a distance of only 13,000 miles (20,921 km). The images *Pioneer 11* took gave us our first close-up look at the planet.

Pioneer 11 carried about 65 pounds (29 kg) of equipment with it, including instruments to receive and transmit data back to Earth, cameras, and instruments to record information about temperature, radiation, and the possible presence of magnetic fields.

*Having completed its mission to Saturn, Voyager 2
moved on to explore another ringed planet, Uranus.
The outer planets are so far away there is a limit to
the amount of information astronomers can gather
from Earth. This durable little probe has helped to
gulf some of the distance.*

Although its primary purpose was to expand our knowledge of Saturn, the durable craft passed an important test along the way. It traveled through the Asteroid Belt unharmed and safely into outer space. By crossing this crucial barrier, *Pioneer 11* revealed to scientists that space travel beyond Mars was possible.

Once in Saturn's neighborhood, the spacecraft continued its valuable work, detecting the presence of two new moons and new rings as well. It also determined that Saturn's largest moon, Titan, was too cold to support life. The temperature recorded there was minus 292 degrees Fahrenheit (-180° C). *Pioneer 11* continued to transmit data back to Earth for almost twenty-two years. The mission lasted until 1995, far longer than expected.

VOYAGER 1 AND VOYAGER 2

In 1977, the United States sent two more probes to the outer planets. *Voyager 1* and *Voyager 2* were identical spacecraft launched one month apart. The information that the *Voyagers* gathered about Saturn taught us more about the planet than we had ever learned before.

Each 1,797-pound (815-kg) craft contained 65,000 parts. They were both equipped with radios, television cameras, and instruments used to conduct experiments. Among them were magnetometers for measuring magnetism, plasma and cosmic ray detectors, and infrared radiometers and spectrometers used to analyze the atmosphere.

Voyager 1 began its trip to Jupiter in September 1977. After completing its work there, it continued on to Saturn, reaching the planet on November 12, 1980. It flew within 78,000 miles (125,525 km) of Saturn and transmitted astonishing images of its rings. For the first time, scientists were able to see a specific part of the ring system. Protruding from the rings are what look like the spokes of a bicycle wheel. These spokes, or radial bands, rotate with the rings.

Voyager 2 was launched in August 1977, one month earlier than

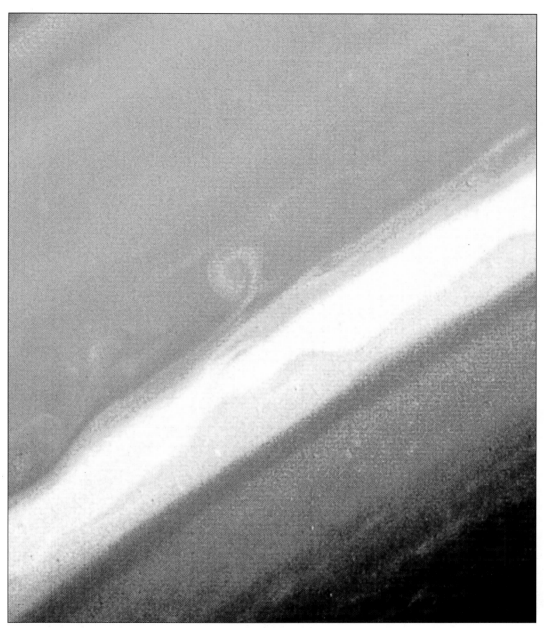

Voyager 2 *studied this curled cloud, part of Saturn's atmosphere, for seven days. Each day on Saturn lasts about 10 hours, 40 minutes.*

MESSAGE IN SPACE

When the *Voyager* spacecrafts were sent into space, each was equipped with a gold-plated record bearing a message recorded especially for the occasion. The greeting is intended for any alien life-form the crafts might encounter. Instructions for playing the record were also included.

Voyager was not the first mission to prepare for contact. *Pioneers 10* and *11* carried simple plaques that identified their origins. But the *Voyager* records take this communication from Earth much farther. A NASA committee headed by astronomer Carl Sagan determined what information the records would bear. Their selections represent the vast diversity of life on Earth.

Each record contains images of planets, art, locations on Earth, wildlife, families, and human activities such as cooking, fishing, and sports. It also has recorded music from many cultures, sounds of nature such as wind and thunder, spoken messages in many languages, and scientific diagrams on subjects such as the Solar System, mathematical and chemical definitions, cell division, and human reproduction.

Kurt Waldheim, then-secretary general of the United Nations, also included his words to our potential neighbors. "I send greetings on behalf of the people of our planet. We step out of our Solar System into the Universe seeking only peace and friendship. We know full well that our planet and all its inhabitants are but a small part of the immense Universe that surrounds us and it is with humility and hope that we take this step."

Carl Sagan was faced with a tough question—what sounds and images should the committee choose to represent the people and places of Earth?

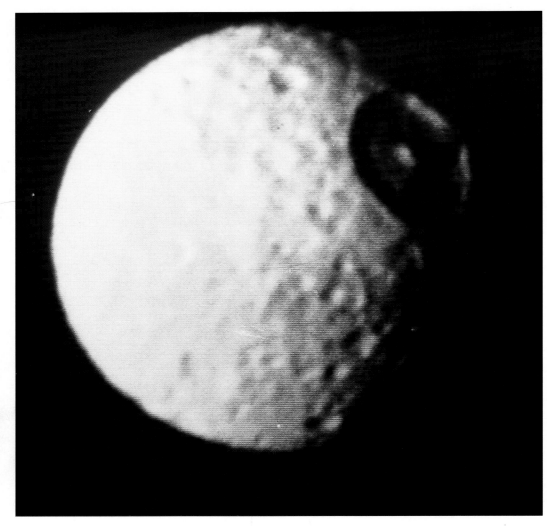

With its distinct crater more than 60 miles (97 km) wide, it is easy to see why Mimas has earned the nicknames Eyeball and Death Star.

Voyager 1. This probe flew by the planet at a distance of 63,000 miles (101,386 km) on August 25, 1981. *Voyager 2* improved our knowledge of Saturn's rings. Composite images taken from both spacecraft dispelled earlier beliefs that Saturn had a few distinct rings. Instead, we now know that Saturn has an entire ring system made up of larger rings and smaller ringlets.

The *Voyager* probes also determined that Titan has a dense atmosphere made up primarily of nitrogen. They sent back images of a dark area near the satellites northern pole. The mission also revealed the presence of the valleys on the moon Enceladus and the moon Mimas's enormous crater. As *Voyager* left the ringed planet and continued on to Uranus, cold, distant Saturn had become a more familiar place.

Saturn's Size
Saturn is a giant. At 75,000 miles (120,698 km) in diameter, more than 760 Earths could fit inside it!

3

RINGS AND THINGS

S aturn is more than 886,000,000 miles (1.43 billion km) from the Sun. With a much greater distance to travel, that means it takes just less than 30 Earth years to make one revolution, or complete trip, around the Sun—29.46 years to be exact. Its period of rotation, however, is not quite as slow. Saturn spins rapidly on its axis, so that a

Voyager 1 provided this view of Saturn's rings and the shadows they cast on the planet's surface. As they sweep around the face of the planet, the rings appear to be transparent.

Saturnian day is only 10 hours, 40 minutes long. Its rapid rotation makes Saturn—like Earth—bulkier at the center and flatter at the poles. So this complex, foreign world does have some things in common with our own.

APPLYING PRESSURE

Although Saturn is the least dense of all the planets, it has a massive atmosphere that applies pressure to the planet's interior. As the pressure pushes down on the gases below, it compresses them, converting gases first to liquids and then to solids.

The center of Saturn is most likely a compressed molten rocky core. Surrounding the inner core is an outer core of liquid metallic hydrogen. Next comes a thick layer of liquid hydrogen. As both the

Inside a gas giant—Saturn's molten rock core (gray) is most likely surrounded by layers of liquid metallic hydrogen, liquid hydrogen, and then hydrogen gas.

temperature and pressure lessen near the surface, the hydrogen becomes gas. *Voyager 1* also determined that, in addition to hydrogen, there is also helium present—about 7 percent—in Saturn's atmosphere.

Saturn's atmosphere consists of distinct layers. The bottom layer is made up of water ice crystals. The middle layer is formed by ammonium hydrosulfide. And the upper layer is ammonia ice crystals. Saturn's lowest temperatures are at the cloud tops, reaching a frigid minus 300 degrees Fahrenheit (-184° C).

The atmosphere appears as colorful bands of yellow, white, and tan. The uppermost clouds are constantly blown by winds swirling up to 1,000 miles per hour (1,609 km/hr). That is almost ten times stronger than hurricane-force winds on Earth.

HOT IN THE CORE

The ringed planet experiences a wide range of temperatures. Imagine that you could begin a space journey right in the core of Saturn. Your spacecraft would have to withstand extreme heat. That is because Saturn's inner and outer cores have temperatures around 21,700 degrees Fahrenheit (12,038° C). As you traveled outward toward the surface and the atmosphere, your craft would have to make a rapid adjustment as the temperature dropped lower and lower.

Many scientists believe that helium is primarily responsible for Saturn's great inner heat. As helium from the atmosphere slowly sinks down into the planet's internal layers, energy is given off. This energy creates heat.

One startling discovery about Saturn was the amount of heat it releases. Since Saturn is so far from its heat source, the Sun, you would expect it to be a cold planet. And as the *Voyager* probes determined, it is. But surprisingly, Saturn gives off more than twice as much heat as it receives from the Sun.

In addition to the notion of sinking helium, there is another theory that could explain why Saturn generates so much heat. When the Solar System was forming— more than 4.5 billion years ago—Saturn was created out of gas, dust, and other material. These materials are thought to be still settling and combining, and this process creates energy, emitting heat.

A MAGNETIC PLACE

Pioneer 11 first confirmed the presence of Saturn's magnetic field and magnetosphere in 1979. But what exactly are they? A magnetic field is the area around a planet where a magnetic force can be detected. And the magnetosphere is the extent of the magnetic field, or how far the particles charged by the magnetic field extend into space.

But where does the magnetic field come from? The liquid metallic hydrogen in Saturn's interior acts as an electrical conductor, generating current. This current is what creates the magnetic field of the planet. The radiation belts that are part of Saturn's magnetosphere reach out more than 1.3 million miles (2.1 million km) from the center of the planet.

RINGS AROUND THE PLANET

The ring system of Saturn is extensive. Stretched from outer edge to outer edge, it spans a distance of about 240,000 miles (386,000 km). That is almost the distance from Earth to the Moon. So what exactly are these wide-reaching rings made of?

Saturn's rings are composed of billions of particles of ice, dust, and rock. These particles vary greatly in size, from dust that is less than half an inch (1 mm) to icy, rocky clumps that are more than half a mile (0.8 km) in length. The rings themselves are less than 1 mile (1.6 km) thick. The ring system surrounds Saturn at its equator.

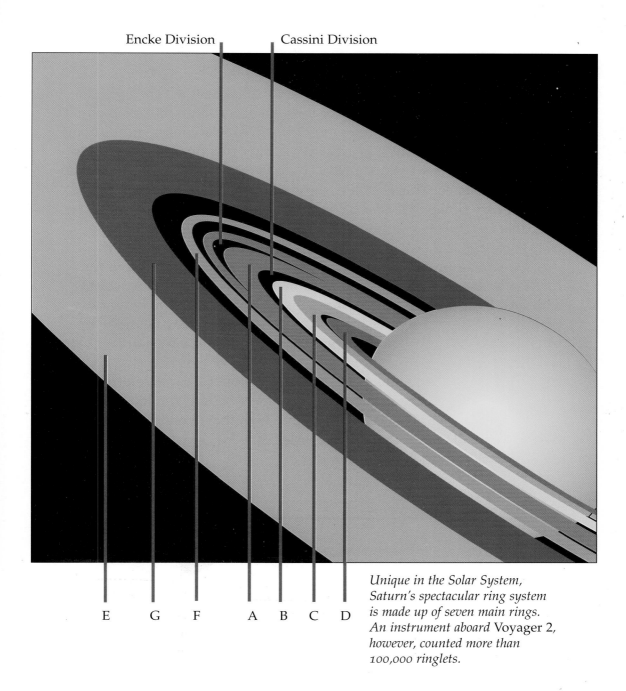

Encke Division   Cassini Division

E   G   F   A   B   C   D

Unique in the Solar System,
Saturn's spectacular ring system
is made up of seven main rings.
An instrument aboard Voyager 2,
however, counted more than
100,000 ringlets.

Therefore, the rings tilt at the same angle as Saturn—26.7 degrees.

At a certain point in Saturn's orbit around the Sun, the rings appear "edge on" from Earth (in other words, imagine turning a coin, so that instead of seeing heads or tails, all you could see was its thin edge). Because the rings are not very thick, they are faint and difficult to view at this time But it is precisely during this time that scientists have a clearer view of Saturn's satellites, as the rings become less of a viewing obstacle. This is called a ring plane crossing, and it happens about every fifteen years, so scientists have plenty of time to prepare in advance. During the 1995 crossing, the Hubble Space Telescope (HST) captured many images of previously unrecorded satellites.

Saturn's ring system includes seven main rings, with thousands of smaller ringlets within each ring. In all, there are perhaps 10,000 ringlets. In addition, some of the rings are marked by the temporary spokes first discovered by *Voyager 1*. These spokes may be made up of dust particles that become charged from time to time by Saturn's magnetic field.

Some of Saturn's rings are separated by open spaces referred to as divisions or gaps. The rings themselves have been given letter names—A through G—in the order in which they were discovered. Here, we'll talk about them in the order that they exist— from the ring closest to Saturn moving outward.

Gravity

The gravitational field of Saturn is about 1.06 times stronger than Earth's. This means that a 100-pound (45-kg) person on Earth would weigh about 106 pounds (48 kg) on Saturn.

The D-ring is the innermost ring and is very faint. Next is the Guerin Division, a narrow gap between the D-ring and the C-ring. It was named for Pierre Guerin who, in 1969, was the first to prove that there was a D-ring. After the Guerin Division is the inner edge of the C-ring. This was the third-known ring and, at about 11,000 miles (17,700 km) wide, is one of the main rings. Next comes the Maxwell Division. Another narrow gap, this one separates the C-ring from the

Computer-processed images from Voyager 2 *show the variations in the colors of Saturn's rings. Astronomers believe that each ring has a different chemical makeup, accounting for the range of colors.*

This simulated image of the A-ring also shows the gap known as the Encke Division.

B-ring. *Voyager* discovered this gap in 1980. It was named after James Maxwell who, in 1856, proposed that Saturn's rings were made up of particles.

Beyond the Maxwell Division is the B-ring. Spokes have been observed in this brightest ring composed of thousands of ringlets. The wide Cassini Division separates the B-ring from the A-ring. This expansive opening spreads out 2,900 miles (4,670 km). Within it, *Voyager* discovered a smaller and separate gap named Huygens after the astronomer Christiaan Huygens.

The A-ring was the first ring to be discovered. Huygens identified it in 1656. This ring also contains many ringlets but is much darker than the B-ring. Toward the outer edge of the A-ring lies a gap named the Encke Division, named after nineteenth-century astronomer Johann Encke. And yet another new gap discovered by *Voyager* in 1980 was named Keeler. James Keeler was the first person to clearly identify the Encke Division as a gap, after Encke had originally described that more inner area as a band.

Following these two gaps are the last three of Saturn's known rings. Though it is very narrow, the F-ring contains a faint ringlet sandwiched between two brighter rings with varied structures of bumps and twists. The G-ring—another *Voyager* discovery—and the E-ring are extremely faint. There is little doubt that *Cassini*, the next planned mission to Saturn will turn up new information about the planet's mysterious system of rings. In fact, the *Cassini* mission will actually travel through the rings.

4

MANY MOONS

Saturn has more moons than any other planet. Although most have rocky cores and icy surfaces in common, each satellite has its own unique characteristics. Many are pockmarked and heavily cratered, some have large smooth plains, while others are covered in ice. Before 1980, scientists already knew of multiple satellites. Interpreting images taken by *Voyager* resulted in the discovery of many new ones. Now we know that there are at least eighteen moons of Saturn.

A GROWING FAMILY

Titan is Saturn's largest moon and can be seen through a telescope from Earth. At 3,190 miles (5,134 km) across, it is larger than the planets Mercury and Pluto. Titan revolves around Saturn once every 16 days. One of the most unusual characteristics of this extremely cold moon is that it has an atmosphere. Clouds in the upper atmosphere give it a fuzzy, orange appearance. We learned from *Voyager* data that Titan's atmosphere is made up primarily of nitrogen. The information gathered also suggests that Titan may have pools of liquid hydrogen on its surface. It may also experience something similar to snowfall. Using infrared imaging, which allows us to "see" beyond the visible

The ringed planet surrounded by six of its moons—Dione, Tethys,
Mimas, Enceladus, Rhea, and Titan (upper left to upper right).

A giant, haze-shrouded moon, Titan's atmosphere is marked by bands of circulating clouds.

light spectrum, the Hubble Space Telescope was able to peer through Titan's clouds for the first time. The images that the telescope provided showed dark patches and one bright area slightly smaller than the size of the United States. These images were used to make a map of Titan.

After Titan, Dione is Saturn's densest moon. It has a rocky core and an icy surface. The terrain on Dione is varied, ranging from smooth areas to lightly cratered plains to areas riddled with huge craters. These craters—many of which are more than 20 miles (32 km) wide—may have been caused by comets or asteroids crashing into the moon's surface. Dione also sports bands of a wispy, unknown substance floating above its surface.

Rhea and Tethys are icy bodies similar to Dione. Rhea is Saturn's second-largest satellite and has the most craters of all its moons. In 1981, *Voyager* recorded a beautiful image of Tethys that gave us new information about this moon. It has a huge trench called Ithaca Chasma about 40 miles (64 km) wide that wraps around nearly 75 percent of the moon. Tethys is also home to the Odysseus Basin, an impact crater nearly 250 miles (402 km) wide. This basin was most likely formed at a time when Tethys was much warmer and could absorb the shock of an object crashing into its surface. If it had been icy and brittle, as it is today, Tethys may have been destroyed by the impact.

Saturn's satellite Phoebe is unusual in that it orbits the planet in a direction opposite that of the other satellites. It is small and red and somewhat circular in shape. It is likely that Phoebe was originally an asteroid that was captured by Saturn.

Hyperion and Iapetus are both somewhat strange. Hyperion is very small and is probably a remnant of a once-larger satellite. Whenever it passes Titan, something unusual starts to happen. The gravity of Titan pulls on Hyperion and causes it to waver and assume an erratic orbit. Iapetus is noted for its two distinct sections. Roughly

Saturn hangs in the distance, above the icy, heavily cratered surface of Rhea.

Some parts of Enceladus show impact craters up to 22 miles (35 km) wide. Other regions are smooth and unbroken. These crater-free areas may indicate that Enceladus has experienced internal melting in the recent past.

Pandora and its partner Prometheus are the shepherd moons of the F-ring. It is their gravity that helps this distant ring keep its shape.

one-half of it is icy and a very bright white. The other half is dark red and thought to be covered in dust.

One of the satellites closest to the surface of Saturn is Mimas. This moon is host to a large crater named Herschel. It appears fairly centered on the moon and covers more than one-third of its diameter. The crater is 6 miles (9.7 km) deep and has a huge mountain protruding from within. Because of this crater, Mimas has been given the nicknames Eyeball and Deathstar.

Enceladus is a bright moon that is probably covered with ice. It is marked by only a few small craters as well as furrows, or grooves, that run for miles. The grooves are most likely faults in the crust of the moon. Because it reflects nearly all the sunlight that hits it, Enceladus practically glows in space.

SHEPHERD SATELLITES

Four of Saturn's moons are considered shepherd satellites. They are called this because they stay close to Saturn's rings and help contain the ring material much like a shepherd keeps a flock of sheep together. How does this work? Ring particles are attracted to the gravity of a satellite that orbits on either the inner or outer edge of a ring. Therefore, the gravity of the satellite helps to establish and define the outer and inner boundaries of a ring. Saturn's known shepherd satellites are Atlas, Pan, Prometheus, and Pandora.

Atlas orbits on the outer edge of Saturn's A-ring. Scientist Richard Terrile identified Atlas after close examination of *Voyager 1* images of the area. Pan is located in the Encke Gap of Saturn's A-ring. The gravitational pull of Pan helps keep the Encke Gap open. Mark Showalter found this small, bright moon after a careful analysis of *Voyager 2* images. He concentrated on an area where scientists believed there might be a small moon.

Prometheus lies on the outer edge of Saturn's F-ring. Pandora is

Saturn's Known Moons

Name	Year of Discovery	Discoverer
Titan	1655	Christiian Huygens
Iapetus	1671	Giovanni Cassini
Rhea	1672	Giovanni Cassini
Dione	1684	Giovanni Cassini
Tethys	1684	Giovanni Cassini
Mimas	1789	William Herschel
Enceladus	1789	William Herschel
Hyperion	1848	W. Bond & W. Lassell
Phoebe	1898	William Pickering
Janus	1966	Audouin Dollfus

Name	Year of Discovery	Discoverer
Epimetheus	1978	T. Fountain & S. Larson
Helene	1980	P. Laques & J. Lecacheux
Telesto	1980	Smith, Larson, & Reitsema
Calypso	1980	Pascu, Seidelman, Baum, & Currie
Atlas	1980	Richard Terrile
Prometheus	1980	S. Collins & D. Carlson
Pandora	1980	S. Collins & D. Carlson
Pan	1990	Mark Showalter

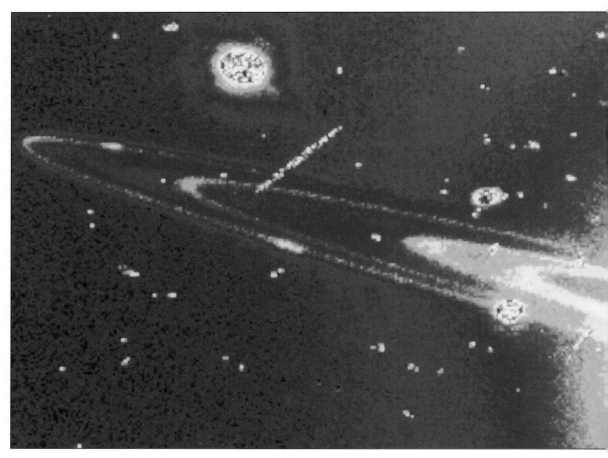

A new moon or a piece of cosmic debris? The Hubble Space Telescope is always on the lookout. This image, taken in 1996, shows the large, bright moon Dione above Saturn's rings. Two smaller moons, Pandora (bottom) and Prometheus (top), are closer to the planet and appear to be touching the F-ring.

located on the inner edge of Saturn's F-ring. Both Prometheus and Pandora are icy bodies with craters on their surfaces. They help keep the F-ring contained and intact.

DID THE HUBBLE SPACE TELESCOPE DISCOVER NEW MOONS?

In 1990, the revolutionary Hubble Space Telescope was launched into space. It circles our planet, viewing objects in space without interference from Earth's atmosphere. Four stories tall and equipped with imaging instruments and cameras, the HST routinely sends remarkable images of bodies in space back to Earth.

In May 1995, the HST sent back images of what appeared to be up to six new moons of Saturn. Scientists at the Lowell Observatory in Flagstaff, Arizona analyzed the images. They calculated that two of these satellites, which had been labeled S/1995 S1 and S/1995 S2, were the previously discovered moons Atlas and Prometheus.

Two others, S/1995 S3 and S/1995 S4, are small bodies with diameters of about 45 miles (72 km) each. S3 is on the outside edge of Saturn's F-ring, and S4 is about 3,700 miles (5,954 km) beyond that. In fact, all of the unnamed satellites orbit near the F-ring. More data gathered in August 1995 from the HST showed that they might actually be pieces of moons that have been blown apart by collisions with other bodies within the rings.

As we continue to analyze images from *Voyager*, as well as plan future missions to Saturn, we may find additional moons of the ringed planet. The next mission to Saturn, *Cassini*, will undoubtedly provide new insight. Included in its plans is a probe that will explore the surface of Titan for the first time.

5

THE CASSINI MISSION

An artist's version of the Cassini spacecraft as it hurtles through space. With both of its engines firing, the craft can alter its speed and direction as needed. This will allow it to be captured by Saturn's gravity more easily.

The immediate future of Saturn's exploration lies with the *Cassini* mission, which is a joint effort between NASA, the European Space Agency (ESA), and the Italian Space Agency.

The spacecraft consists of the *Cassini* orbiter and the Huygens probe. The orbiter is approximately 22 feet (6.7 m) tall and 4 feet (1.2 m) wide. It weighs about 4,750 pounds (2,155 kg) and carries another 1,500 pounds (680 kg) of scientific equipment, including the probe.

The *Cassini* spacecraft was launched on October 15, 1997, from Cape Canaveral, Florida. It embarked on a seven-year journey and is expected to reach Saturn's orbit in July 2004, beginning an exciting new chapter in the history of Saturn's exploration. By April 2000 it had already crossed the unpredictable Asteroid Belt, which lies between Mars and Jupiter, making it only the seventh craft to pass this strategic point in our Solar System. By October 2000, *Cassini* was even closer to its goal and sent back images of Jupiter as it neared that planet.

*The **Cassini** orbiter will not just approach Saturn's rings. If all goes well, it will fly right through them.*

Because Saturn is so far from the Sun, the Cassini craft is not able to use solar power as an energy source. Instead, it is equipped with three radioisotope thermoelectric generators (RTGs) that provide its electrical power. RTGs work by converting the energy that is released by naturally decaying plutonium. Although they have been subject to controversy, RTGs have been used successfully on previous missions.

One of the exciting things about the *Cassini* mission is that the spacecraft is going to fly right through Saturn's ring gaps in order to

get close enough to do its work. In 1995, the Hubble Space Telescope was able to learn a bit more about the thickness of the rings and the gaps between them. Scientists used this information to help them plan the maneuver through a gap between the F-ring and the G-ring, as well as through the E-ring. About traveling through Saturn's rings, Jet Propulsion Laboratory (JPL) scientist Linda Horn said, "We're going in awfully close . . . so the more we know about the boundaries of the rings, the more confident we'll be."

A few months after the mission reaches Saturn, the Huygens probe will venture out on its own. It will be released from the orbiter to travel to Titan and gather information. *Cassini* will orbit the ringed planet for four years, conducting experiments and gathering data on many elements of the Saturnian system. Cameras on board the orbiter will take a half million images. *Cassini*'s main objectives are to complete in-depth studies of the planet itself, as well as of its atmosphere, rings, moons, and magnetosphere.

THE HUYGENS PROBE

The Huygens probe is a robotic laboratory that was engineered by the ESA. Its main goal is to travel through the atmosphere of Titan and land on its surface in an effort to learn more about Saturn's largest moon. If all goes according to plan, the probe will be released from the orbiter in November 2004. It is an historic mission as it is the first time a space probe will attempt to land on the surface of another planet's satellite.

Once Huygens is separated from the orbiter, five powerful batteries will fuel the probe. Huygens will coast toward the moon for twenty-two days before reaching its atmosphere. The descent through Titan's atmosphere will take about two and a half hours and will be executed with a system of three parachutes. The images that the Hubble Space Telescope has already taken of Titan helped scientists

After leaving the Cassini *orbiter, the Huygens probe will drop down through Titan's atmosphere, taking measurements along the way. If it survives the high-speed landing, it will continue its work from the surface. Data from the impact will help scientists determine the makeup of Titan's surface. Is it rock, ice, slush, or something else altogether?*

choose landing targets and calculate the speed of winds through which the parachutes will sail.

The Huygens probe system includes Probe Support Equipment (PSE) that stays attached to the orbiter. The PSE will track the probe. It will also receive data information from the probe, which will then be transferred to the orbiter, and finally transmitted back to Earth. Huygens will begin sending information as soon as it begins its descent, greatly increasing our knowledge about Titan's atmosphere.

Huygens is equipped with six main scientific instruments. Some are specially designed to study Titan's atmosphere during the probe's descent. The Doppler Wind Experiment (DWE) will measure the motion of the probe caused by winds in the atmosphere. The Aerosol Collector and Pyrolyser (ACP) will collect, heat, and examine aerosol particles from the atmosphere.

Another device, the Huygens Atmospheric Structure Instrument (HASI), will concentrate on measuring some of the physical properties of Titan's atmosphere such as temperature and pressure. The HASI will also analyze surface conditions. Likewise, the Descent Imager/Spectral Radiometer (DISR) will test both atmospheric and surface conditions. The Gas Chromatograph Mass Spectrometer (GCMS) will analyze a variety of gases in Titan's atmosphere, as well as gases on the moon's surface, to determine their chemical composition.

Once Huygens has landed on Titan's surface, it will have only enough power left to function for perhaps another thirty minutes. During this time, the Surface-Science Package (SSP) will be put into action. Using the SSP, Huygens will perform a series of experiments in rapid succession, testing the makeup of the surface at the landing site and taking full advantage of every precious second on Titan.

Why do all of this for a group of experiments that will last only a matter of hours? Titan's atmosphere is nitrogen based, similar to Earth's, and could have the characteristics needed for that moon to

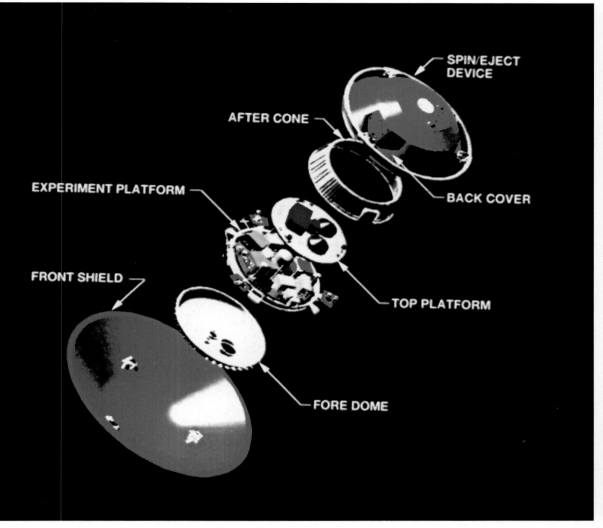

A well-equipped robotic laboratory, the Huygens probe will record more than 1,100 images of Titan.

evolve into a life-sustaining body. Scientists are hopeful that exploring Titan more thoroughly will offer some clues into the ancient, prelife history of Earth and teach us more about the evolution of our own planet.

CASSINI SCIENCE

Like the Huygens probe, the *Cassini* orbiter is also equipped with a wealth of instruments that will image, record, and analyze information. Some will use infrared and ultraviolet imaging techniques, allowing us to see beyond the visible light spectrum.

These instruments will help researchers study the atmospheric temperature, the gases in the atmosphere, and how neutral and charged particles are distributed within the magnetosphere. They will also aid in determining wind and other atmospheric patterns, what the surfaces of Saturn's icy satellites are made of, and whether lightning occurs on either Saturn or Titan.

The Composite Infrared Spectrometer (CIRS) will perform a variety of functions such as mapping surface temperatures of Titan and the makeup of Saturn's rings and moons. And the Ultraviolet Imaging Spectrograph (UVIS) will measure the ultraviolet light given off or reflected by elements of the Saturn system. The UVIS will use this information to learn more about Titan's and Saturn's atmosphere and the structure of its rings.

One of the instruments that will be used to send back images is called the Imaging Science Subsystem (ISS). This uses both a narrow-angle and a wide-angle camera and will map both Saturn's and Titan's atmosphere. The ISS will also study the gravitational relationship between the rings and satellites and determine more precisely the thickness and composition of the rings.

Other instruments, such as the Cassini Radar and the Radio Science Subsystem (RSS), will use microwave technology to gather

Astronomers want a clearer picture of Titan's surface. Some theorize this largest moon of Saturn is covered with oceans of methane gas. Cassini's radar imaging will hopefully provide some answers.

THE RTG CONTROVERSY

RTGs are power supply systems for spacecraft that use plutonium, a radioactive material, as a heat source. The RTG captures the heat and converts it into electrical power. Around the time of the *Cassini* launch, anti-nuclear activists protested the use of RTGs. They feared that if the spacecraft suffered an accident, the radioactive material could be released and be harmful to people.

NASA responded by explaining that, unlike a nuclear reactor, an RTG is stable and could never explode. Furthermore, NASA said that even in the highly unlikely event that some of the plutonium was ever released into the atmosphere it would be spread over the globe and most of it would never be absorbed by people.

Although the *Cassini* engineers and scientists took every precaution, some people still believe that it is a mistake to take any risk, no matter how small, of contaminating people's health. The good news is that although there have been many past missions that have used this technology, and even some RTGs that did break apart, there has never been any resulting contamination.

information. Cassini Radar will look through Titan's thick clouds in search of oceans and any other solid land forms on the satellite. It will also possibly try to collect data on other moons. The RSS will send out and receive microwaves in order to see how the varying sizes of particles are spread throughout Saturn's rings. This instrument will also gather information concerning the temperatures, makeup, and pressures of Saturn's and Titan's atmospheres.

Still other instruments aboard *Cassini* will focus on learning more about the particles and fields in the Saturn system. The Cassini Plasma Spectrometer (CAPS) and the Ion and Neutral Mass Spectrometer (lNMS) will measure both charged and neutral particles in Saturn's magnetosphere, as well as in Titan's atmosphere. From these measurements, CAPS and INMS will determine the temperatures and densities of the particles, as well as how they move.

The Cosmic Dust Analyzer (CDA) will study the physical and chemical properties of Saturn's dust. It will help scientists learn more

---

Saturn Fact Sheet

Mean distance from Sun: 886,000,000 miles (1.43 km)
Diameter: 75,000 miles (120,698 km)
Mean atmospheric temperatures: They range from -130°C
     (-202°F) to -190°C (-310°F)
Surface gravity: 1.06
Period of revolution (year): 29.46
Period of rotation (day): 10 hours, 40 minutes
Number of satellites: 18 confirmed moons

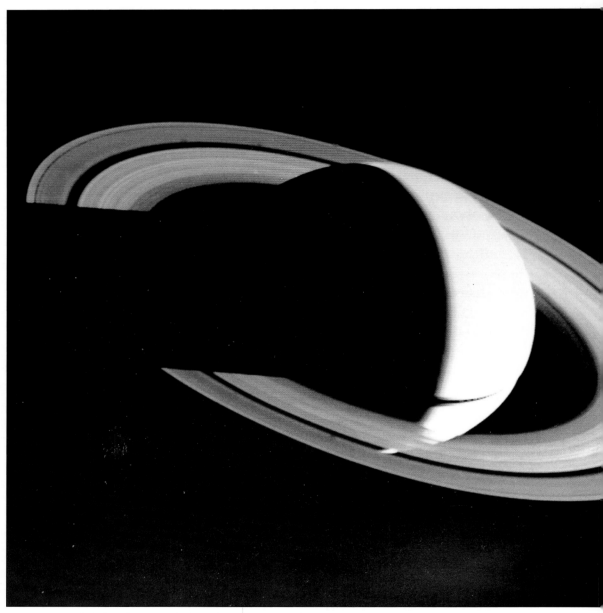

Once called the planet with "ears," generations of scientists have slowly brought Saturn out of the shadows. One day perhaps it will be a planet as familiar to us as our own.

about the chemical makeup of ring particles and how the icy satellites are a source of ring particles. It will also aid in the study of how particles first formed the E-ring and of how dust affects the magnetosphere.

For an in-depth study of Saturn's magnetosphere, *Cassini* is carrying the Dual Technique Magnetometer (MAG). MAG will allow scientists to create a three-dimensional model of the magnetosphere. It will also study how satellites, ring particles, and dust interact with the magnetosphere.

These and other studies will help create an enormous database of information about the Saturnian system that scientists will most likely analyze for years. The information the mission gathers will provide valuable clues as we continue to solve some of Saturn's deepest mysteries. And, of course, there is no telling what surprising new wonders of the ringed planet *Cassini* will discover.

GLOSSARY

asteroid an object made of rock or metal that revolves around the Sun; a large meteoroid

axis a real or imaginary straight line around which a planet rotates

chasm a large canyon

comet a heavenly body that travels around the Sun and is composed of ice, frozen gases, and dust that has a long tail

cosmic rays high-frequency rays that are made up mainly of positively charged particles

electromagnetic spectrum a grouping of various forms of radiation, arranged from longest wavelength to shortest—radio waves, microwaves, infrared, visible light, ultraviolet, X-rays, and gamma rays

infrared the red end of the electromagnetic spectrum not visible to the naked eye

magnetic field the area surrounding a planet where a magnetic force can be detected

magnetometer an instrument used to detect the presence of a metallic object or to measure the intensity of a magnetic field

magnetosphere the extent of the magnetic field

radiometer a device used to detect and measure the strength of thermal energy, particularly infrared radiation

revolution the movement of the planets in an orbit around the Sun

rotation the act of turning on an axis

satellite an object, either natural or human-made, that revolves around a planet

shepherd satellite a satellite that helps contain or shape planetary ring through the force of its gravity

spectrometer an instrument that measures the wavelengths of light or determines the chemical makeup of a sample

ultraviolet the violet end of the electromagnetic spectrum that is not visible to the naked eye

FIND OUT MORE

BOOKS FOR YOUNG READERS

Branley, Franklyn. *Saturn: The Spectacular Planet*. New York: Crowell, 1983.

———. *The Sun and the Solar System*. New York: Twenty-First Century Books, 1996.

Clay, Rebecca. *Space Travel and Exploration*. New York: Twenty-First Century Books, 1996.

Jackson, Francine. *Outer Space: The Outer Planets* vol. 4. Danbury, CT: Grolier, 1998.

Landau, Elaine. *Saturn*. New York: Franklin Watts, 1999.

Redfern, Martin. *The Kingfisher Young People's Book of Space*. New York: Kingfisher, 1998.

Solomon, Maury. *An Album of Voyager*. New York: Franklin Watts, 1990.

VanCleave, Janice. *Astronomy for Every Kid*. New York: John Wiley & Sons, 1991.

OTHER BOOKS

Cooper, Henry S. F. Jr. *Imaging Saturn: The Voyager Flights to Saturn.* New York: Holt, Rinehart & Winston, 1982.

Fimmel, Richard O., Swindell, William, and Burgess, Eric. *Pioneer Odyssey.*

Washington, D.C.: National Aeronautics and Space Administration, 1977.

Fradin, Dennis Brindell. *The Planet Hunters.* New York: Margaret McElderry Books, 1997.

Henbest, Nigel. *The Planets: A Guided Tour of our Solar System through the Eyes of America's Space Probes.* New York: Viking, 1992.

Littmann, Mark. *Planets Beyond: Discovering the Outer Solar System.* New York: John Wiley & Sons, 1988.

Miller, Ron, and Hartmann, William K. *The Grand Tour: A Traveler's Guide to the Solar System.* New York: Workman, 1981.

Ridpath, Ian. *Eyewitness Handbooks: Stars and Planets.* New York: Dorling Kindersley, 1998.

Watters, Thomas R. *Planets: A Smithsonian Guide.* New York: Macmillan, 1995.

Wilson, Colin. *Starseekers.* New York: Doubleday, 1980.

To learn more about the *Cassini* mission to Saturn, visit the following website:

To learn more about *Cassini*'s Huygens probe, visit:
http://sci.esa.int/huygens/

For a site that has a lot of great links and information about Saturn's rings and moons, visit:
http://www.jpl.nasa.gov/cassini/Kids/stories/saturn.html

For a library of activities for kids and teachers that are space-related go to the Jet Propulsion Laboratory's Project SPACE web site at:
http://learn.jpl.nasa.gov/projectspacef/activity.html

For general Saturn sites, visit any of the following:
http://starchild.gsfc.nasa.gov/docs/StarChild/solar_system-level2/saturn.html

http://csep10.phys.utk.edu/astrlbl/lect/saturn/features.html

http://micro.ee.nthu.edu.tw/~u840912/node13.html

About the Author

Tanya Lee Stone is a former editor of children's books who now writes full time. She holds a master's degree in science education and is the author of more than a dozen books for kids, including *Saturn, Rosie O'Donnell: America's Favorite Grownup Kid*, and *The Great Depression and World War II*. She lives in Burlington, Vermont, with her husband, Alan, and her son, Jacob.

INDEX